General Classification Handbook for Floodplain Vegetation in Large River Systems

By Jennifer J. Dieck and Larry R. Robinson

Chapter 1 of
Book 2, Collection of Environmental Data, Section A, Biological Science

Techniques and Methods 2 A–1

U.S. Department of the Interior
U.S. Geological Survey

U.S. Department of the Interior
Gale A. Norton, Secretary

U.S. Geological Survey
Charles G. Groat, Director

U.S. Geological Survey, Reston, Virginia: 2004

For sale by U.S. Geological Survey, Information Services
Box 25286, Denver Federal Center
Denver, CO 80225

For more information about the USGS and its products:
Telephone: 1-888-ASK-USGS
World Wide Web: *http://www.usgs.gov/*

Suggested citation:
Dieck, J. J., and Robinson, L. R., 2004, Techniques and Methods Book 2, Collection of Environmental Data, Section A, Biological Science, Chapter 1, General classification handbook for floodplain vegetation in large river systems: U.S. Geological Survey, Techniques and Methods 2 A–1, 52 p.

Library of Congress Cataloging-in-Publication Data

Dieck, Jennifer J.
General classification handbook for floodplain vegetation in large river systems / by Jennifer J. Dieck and Larry R. Robinson.
p. cm. -- (Techniques and methods ; bk. 2, A-1)
Includes bibliographical references and index.
1. Marsh plants. 2. Floodplain plants. 3. Vegetation classification. I. Robinson, Larry R. II. Title. III. Series.
QK938.M3D54 2004
581.7'68--dc22

2004011419

Contents

Figures

Tables

General Classification Handbook for Floodplain Vegetation in Large River Systems

By Jennifer J. Dieck and Larry R. Robinson

Abstract

This handbook describes the Long Term Resource Monitoring Program (LTRMP) General Wetland Vegetation Classification System developed as part of a multi-state and Federal partnership for the Upper Mississippi River System. This classification system consists of 31 general classes and has been used to create systemic land cover/land use maps throughout this diverse river system. In addition, it describes the evolution of the General Wetland Vegetation Classification System, discusses the process of creating a map, and describes each of the 31 general classes in detail. This handbook also acts as a pictorial guide to illustrate each of the general classes as they may appear in the field, as well as on color-infrared aerial photographs.

Introduction

Vegetation mapping can be one of the most important tools used in vegetation science and landscape ecology (Zonneveld, 1988). It creates an inventory of existing vegetation types, along with their location and geographical distribution at a particular moment in time. Changes in vegetation often occur more quickly and distinctly than in other ecological variables, making vegetation a sensitive indicator of environmental changes (Zonneveld, 1988). Because vegetation is such an important indicator in the environment, scientists from many different disciplines are interested in vegetation maps. These maps can be used to analyze the relations between vegetation types at a particular site or as a reference for observing and measuring change over time. They can also be extremely important as a basis for future land use planning.

This handbook describes a wetland vegetation classification system developed for large river floodplains in the Upper Midwest and how it can be used in the interpretation of aerial photographs to create vegetation maps. The classification system consists of 31 general classes and primarily has been used to interpret floodplain vegetation of the Upper Mississippi and Illinois River Systems from color-infrared (CIR) aerial photographs. We provide photographs and descriptions

of each of the 31 general classes. Also included are 2- x 2-inch CIR images of each general class extracted from interpreted aerial photographs, along with a description of the signature (or photographic appearance).

Development of the Classification System

Vegetation mapping is a critical element of the Long Term Resource Monitoring Program (LTRMP), a multi-state and Federal partnership created under the Water Resources Development Act of 1986. The mission of the LTRMP is to provide decision makers with information to help facilitate maintenance of the ecosystem and navigation values of this river. The long-term goals of the program are to understand the system, determine resource trends and impacts, develop management alternatives, and manage information. Vital to the LTRMP is the ability to accurately and efficiently map floodplain land cover/land use (LCU) information. These data provide the framework for geographic information system (GIS) analyses and are a crucial component of the Habitat Needs Assessment project (U.S. Army Corps of Engineers, 2000), an ongoing evaluation of the existing habitat conditions throughout the UMRS that guides the selection, design, and evaluation of habitat rehabilitation and enhancement projects.

Two different classification schemes have been used to create vegetation maps. One is a detailed classification system that includes 151 vegetation classes and provides genus-level information. This classification system originated in the mid-1970s as part of a Great River Environment Action Team (GREAT) study designed to look at the use of CIR aerial photography for a habitat analysis of the Upper Mississippi River floodplain (Hagen and others, 1977). A second, and more general, vegetation classification system is the focus of this handbook. In this system, there are 31 general classes. This system, referred to as the General Wetland Vegetation Classification System, complements the more detailed genus-level classification. The combination of these two approaches creates a two-tiered hierarchical system that can easily be adapted to different scales and needs—a hydrology-based, genus-level classification for focused studies that also collapses into broader plant categories while maintaining the hydrologic distinctiveness of the detailed classes.

Relation to Other Classification Systems

Several vegetation classification systems have been widely used in describing land cover patterns. These include Anderson and others (1976), Cowardin and others (1979), and the National Vegetation Classification System (NVCS; Federal Geographic Data Committee, 1997; Grossman and others, 1998). The Anderson Classification System was developed for use with remote sensing systems in the 1970s and is made up of a two-level hierarchy (Anderson and others, 1976). The Cowardin Classification System places ecologically similar habitats into a more complex hierarchal system that includes several layers of detail for wetland classification (Cowardin and others, 1979). Lastly, the NVCS was developed for use in conservation planning and biodiversity protection, as well as for the basic understanding of ecological patterns (Federal Geographic Data Committee, 1997; Grossman and others, 1998). The NVCS is also hierarchical and combines physiognomy and floristics. The General Wetland Vegetation Classification System includes a crosswalk to the formation-level of the NVCS in its attribute table. This allows resource managers, researchers, and analysts to view and analyze the data at the 31-class level or the NVCS-equivalent.

The General Wetland Vegetation Classification System described in this handbook is most similar to the Cowardin Classification System. Both rely heavily on hydrologic regime as the fundamental basis, and the regimes used to classify our system are derived from Cowardin and others (1979). Accordingly, each of the 31 classes in the General Wetland Vegetation Classification System is associated with one of the six following hydrologic regimes:

Permanently Flooded—Water present all year round

Semipermanently Flooded—Water present throughout the growing season, except in periods of extreme drought

Seasonally Flooded—Water present for most of the growing season

Temporarily Flooded—Water only present early in the growing season

Saturated Soil—Soils that are saturated with water during the growing season

Infrequently Flooded—Water rarely present

Discussion

The UMRS has been classified and mapped twice as part of the LTRMP: in 1989 based on 1:15,840-scale aerial photography that was interpreted using a genus-level system and in 2000 based on aerial photography collected at the 1:24,000-scale using the coarser, 31-class, General Wetland Vegetation Classification System. Based on our experience in classifying and mapping the 1989 aerial photography, we redesigned the classification system and the mapping protocols when we undertook the 2000 mapping effort. This redesign maintains critical vegetation information, but creates a more timely product.

When the 1989 mapping began, resource managers wanted to know what species, or mix of species, were present, as well as a sense of the associated hydrologic regimes. To accommodate these desires, we began with a vegetation classification system that combined genus and genus mixture information with the National Wetland Inventory (NWI) vegetation classification naming conventions. The NWI classification includes hydrologic information and broad vegetation categories in its class descriptions, so the blending of these two systems provided much of the information managers were wanting. Unfortunately, this combination made the classification cumbersome and difficult to consistently apply. The NWI component was subsequently dropped, and the 1989 aerial photographs were interpreted only with the genus-level information.

Often, a 1-acre minimum mapping unit (MMU) is used when interpreting 1:15,840-scale aerial photographs, but for the 1989 aerial photography, all distinctive features that could be identified and delineated were mapped. This process gave managers and analysts an unprecedented level of habitat information, but it also resulted in long photointerpretation times and complex photo overlays. This effort was compounded by the time required for quality control and conversion to a digital format. The completed data set presented a detailed snapshot of floodplain habitat, but the process took more than a decade to complete at an annual cost exceeding $100,000.

A new classification system was developed for use with the 2000 aerial photography that could be used to map floodplain vegetation much more rapidly and efficiently. When the classification system was revised, genus, genus-dominance for mixed classes, and hydrology became the primary factors in determining the plant categories. This created a new, detailed 151-class classification system that is typically used for focused studies.

Furthermore, the detailed classes are able to collapse into broader, but still ecologically useful categories based on hydrology. These broader categories are the 31 general classes in the General Wetland Vegetation Classification System. This classification system is designed for use in systemic studies, where aerial photographs are often taken at a smaller scale and a larger 1-ha MMU is used for photointerpretation. This process was used to interpret and map the 2000 aerial photography at about half the cost and in less than half the time of the 1989 aerial photography.

Our experience indicates that there is no perfect method for mapping vegetation. No matter how detailed or general a classification system is, delineating diverse habitats with limited-resolution satellite or aerial imagery will never be free of some subjectivity. A classification system that is flexible, easily updated, and applicable at various scales will have the greatest long-term utility. Our experience also suggests three elements are critical when developing a classification system: (1) the management needs of decision makers (how detailed the vegetation needs to be described), (2) the funding available, and (3) how quickly the data are needed.

These factors will determine the scale of the photography, how habitats should be classified, and the funding and personnel needed to complete the mapping within a given time frame. A carefully designed and implemented vegetation mapping program can be one of the most useful tools a resource manager has for making decisions today and in the future. If possible, the classification system used should also be compatible with other vegetation classification systems to ensure that its scope is extended, as well as its longevity.

The appendixes of this handbook (Appendixes 1–4) provide more detailed information about the classification and mapping of floodplain vegetation using the General Wetland Vegetation Classification System. Appendix 1 describes the process of creating a LCU map. This process includes aerial photograph acquisition, field reconnaissance, photointerpretation, conversion of photointerpreted data to a digital format, and accuracy assessment. Appendix 2 describes each of the 31 general classes as they appear in their environment, as they relate to a hydrologic regime, and as they appear to the photointerpreter on the aerial photograph. These descriptions are complemented with photographic representations of each class in the field, as well as 2- x 2-inch CIR images extracted from interpreted aerial photographs. Appendix 3 provides a classification key used during field reconnaissance to classify land features or vegetation types into the General Wetland Vegetation Classification System. Lastly, Appendix 4 provides a list of the predominant species and common 31 general classes associated with the genera referred to in this handbook.

Acknowledgments

We thank Erin Hoy who completed most of the photointerpretation work involving the CIR aerial photographs for this handbook. Thanks to Heidi Langrehr for preparing the vegetation key and for offering additional comments. Thanks also to Sara Lubinski, Kevin Hop, Steve Zigler, Pat Heglund, and Kirk Lohman for reviewing the handbook. All photographs were taken by staff at the Upper Midwest Environmental Sciences Center. This work was completed as a part of the LTRMP (LTRMP Technical Report 2004-T003).

References Cited

Anderson, J.R., Hardy, E.E., and Roach, J.T., 1976, Land use and land cover classification system for use with remote sensing data: U.S. Geological Survey Professional Paper 964, A revision of the land use classification system as presented in U.S., Geological Circular 671, U.S. Government Printing Office, Washington, D.C., 26 p.

Cowardin, L., Carter, V., Golet, F., and LaRoe, E., 1979, Classification of wetlands and deepwater habitats of the United States: U.S. Department of the Interior, Fish and Wildlife Service, Washington, D.C., 103 p.

Environmental Systems Research Institute, National Center of Geographic Information and Analysis, and The Nature Conservancy, 1994, NBS/NPS Vegetation Mapping Program: Accuracy assessment procedures, Prepared for the U.S. Department of the Interior, National Biological Survey and National Park Service, Washington, D.C., 107 p.

Federal Geographic Data Committee, 1997, Vegetation classification standard, FGDC-STD-005, Web address: *http://www.fgdc.gov/standards/documents/standards/vegetation.*

Grossman, D.H., Faber-Langendoen, D, Weakley, A.S., Anderson, M., Bourgeron, P., Crawford, R., Goodin, K., Landaal, S., Metzler, K., Patterson, K.D., Pyne, M., Reid, M., and Sneddon, L., 1998, International classification of ecological communities: terrestrial vegetation of the United States, Volume I, The National Vegetation Classification System: development, status, and applications, The Nature Conservancy, Arlington, Virginia, 89 p. + Appendixes A–E.

Hagen, R.T., Werth, L.F., and Meyer, M.P., 1977, Upper Mississippi River Habitat Inventory, A report of research by the Remote Sensing Laboratory of the College of Forestry and the Agricultural Experiment Station, Institute of Agriculture, Forestry and Home Economics, University of Minnesota, St. Paul, Minnesota, IAFHE RSL Research Report 77–5, 18 p.

Owens, T., and Hop, K.D., 1995, Long Term Resource Monitoring Program standard operating procedures: Field station photointerpretation. National Biological Service, Environmental Management Technical Center, Onalaska,Wisconsin, August 1995, LTRMP 95–P008–2, 13 p. + Appendixes A–E.

U.S. Army Corps of Engineers, 2000, Upper Mississippi River System Habitat Needs Assessment: Summary report 2000, U.S. Army Corps of Engineers, St. Louis District, St. Louis, Missouri, 53 p.

Zonneveld, I.S., 1988, Introduction to the application of vegetation maps, *in* Kuchler, A.W., and Zonneveld, I.S., eds., Vegetation mapping, Kluwer Academic Publishers, Norwell, Massachusetts, p. 487–490.

Appendix 1.   Using the General Wetland Vegetation Classification System to Describe and Monitor Land Cover

Appendixes 1 and 2 describe the use of the General Wetland Vegetation Classification System. Appendix 1 provides a brief summary of the methods used to interpret aerial photographs and classify wetland vegetation. The steps in this process include acquiring aerial photographs, conducting field reconnaissance to verify vegetation signatures, delineating vegetation types on photographs (photointerpretation), converting interpreted aerial photo overlays to digital formats (automation), and finally assessing the accuracy of the vegetation map based on field comparisons. Appendix 2 provides a detailed description of each of the 31 general classes used in the General Wetland Vegetation Classification System.

1.   Acquisition of Aerial Photographs

Aerial photography is generally acquired in late summer (late July to early September) when aquatic vegetation is at peak biomass and when water levels are typically stable. The scale at which the photography is taken is dependent upon resolution needs and cost limitations. The General Wetland Vegetation Classification System is primarily used to interpret smaller-scale photography, whereas the detailed classification system (or genus-level classification system) is primarily used to interpret larger-scale photography. A comparison of small- and large-scale 9- x 9-inch aerial photographs are shown in fig. 1–1.

Color-infrared (CIR) aerial photographs are preferred over true-color aerial photographs because reflectance by vegetation is directly related to chlorophyll content, and the more vigorous the growth, the greater the reflectance. This helps the photointerpreter to better distinguish between plant and community types. However, CIR photography does not penetrate water well, and submersed vegetation at low densities may be difficult to identify.

A flight line index is prepared before the CIR aerial photographs are collected (fig. 1–2). The flight line index is created with a 30% side lap (between flight lines) and a 60% end lap (within flight lines) between photographs. This creates a stereoscopic coverage that allows the photointerpreter to perceive depth on the overlapping portions of the photographs (fig. 1–3).

2.   Field Reconnaissance

Before the CIR photographs are interpreted, field reconnaissance is performed (fig. 1–4). Questionable areas on

Smaller-scale photograph
Scale - 1:24,000
Area = 11.6 sq. miles
Date of photograph: 09/00

Larger-scale photograph
Scale - 1:15,840
Area = 4.5 sq. miles
Date of photograph: 08/00

Figure 1–1. Comparison of small- and large-scale 9- x 9-inch aerial photographs.

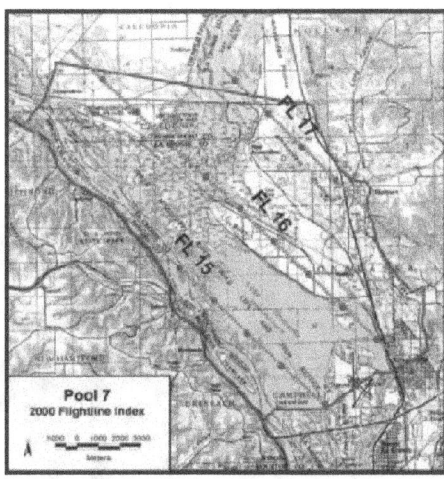

Figure 1–2. Example of the 2000 flight line index for Pool 7, Mississippi River.

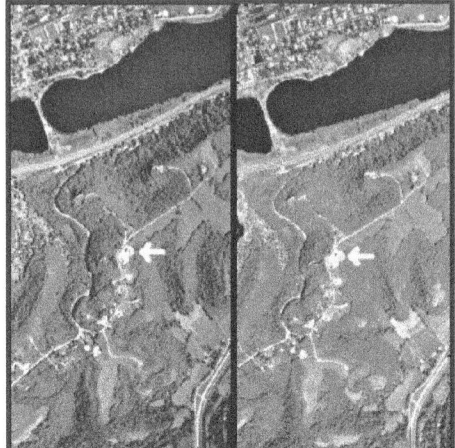

Figure 1–3. An example of two overlapping photographs. (If you relax your eyes and fuse the arrows above, you should be able to see in 3-D.)

the photographs are visited and the plants or land features observed in the area are recorded for reference on a clear sleeve registered to the aerial photograph. This procedure verifies vegetation signatures on the photograph with those on the ground. A vegetation signature consists of several factors, such as color/tone, texture, pattern, shape, size, and location of the vegetation type.

During the field reconnaissance process, a key may be used to help classify a particular land feature, vegetation type, or combination of vegetation types into the 31 general classes (Appendix 3). Note that the key gives examples of some, but not all, of the predominant vegetation types in the Upper Mississippi and Illinois River Systems. The user of the key may need to extrapolate from the examples given and link certain species (e.g., *Polygonum*) to a similar hydrology. The user of the key will also look at the actual percent of the relative cover for each vegetation type. This will determine which of the 31 general classes best describes the area observed. For example, the vegetation in an area may have a total cover of 90%, with a relative cover of 60% *Sagittaria* and 30% *Scirpus*. The dominant vegetation type will determine which of the 31 general classes would best describe the area. In this example,

Deep Marsh Perennial would best describe the area observed because *Sagittaria* is the dominant vegetation type. A more specific list of the predominant species and common 31 general classes associated with the genera referred to in this handbook are in Appendix 4. Once all questionable areas are investigated, the photointerpretation process proceeds.

3. Photointerpretation

Photointerpretation of CIR aerial photographs is performed with a stereoscope. The photographs are interpreted following photointerpretation rules and procedures as established by Owens and Hop (1995). Before photointerpretation begins, a minimum mapping unit (MMU; smallest unit mapped) is determined. The MMU is dependent upon the resource manager's needs, the resolution of the photography, and the cost of the project. For example, CIR aerial photographs taken at a smaller scale such as 1:24,000 generally allow for a MMU of 1 ha (2.5 acres; fig. 1–5). Furthermore, because of the increased resolution of larger scale photography, CIR aerial photographs taken at a scale of 1:15,840 generally allow for a MMU of 0.4 ha (1 acre).

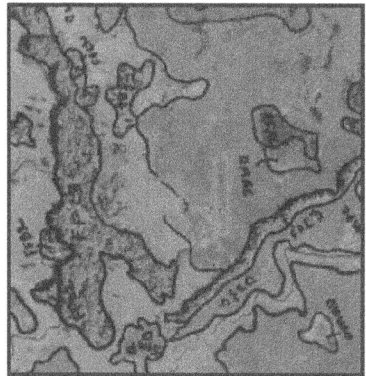

Figure 1–5. The red boxes depict two different shapes of the minimum mapping unit (1 ha or 2.5 acres) used when photointerpreting 1:24,000-scale photography.

During the photointerpretation process, vegetation polygons are delineated on overlays to the 31 general class level and labeled with an attribute, or map code (fig. 1–6). The map

Figure 1–4. Examples of field reconnaissance.

Figure 1–6. (*left*) A Bausch and Lomb Zoom 240 Stereoscope along with an example of completed line work and attributes (*right*).

code represents a vegetation type or land feature, followed by modifiers when applicable. The modifiers, which represent density and height, were developed with respect to what the photointerpreter can reliably identify, and what may be ecologically meaningful to the user. The 31 general classes or map codes along with their respective modifiers are shown in Table 1–1.

Density is determined in each polygon according to the relative cover of that polygon and by life form, with the taller life form taking precedence when more than one vegetative layer exists. For example, if delineating a polygon that contains both rooted-floating vegetation and submersed vegetation, the density would only apply to the rooted-floating vegetation because it is the taller life form. Another example

Table 1-1. The 31 general map classes together with their respective codes, hydrologic regimes, and modifiers. Density and height modifiers designated by an X indicate that they apply to that map class.

Map class	Map code	Hydrologic regime	Density	Height
Open Water	OW	Permanently Flooded		
Submersed Vegetation	SV	Permanently Flooded	X	
Rooted-Floating Aquatics	RFA	Permanently Flooded	X	
Deep Marsh Annual	DMA	Semipermanently Flooded	X	
Deep Marsh Perennial	DMP	Semipermanently Flooded	X	
Shallow Marsh Annual	SMA	Seasonally Flooded	X	
Shallow Marsh Perennial	SMP	Seasonally Flooded	X	
Sedge Meadow	SM	Temporarily Flooded	X	
Wet Meadow	WM	Saturated Soil	X	
Deep Marsh Shrub	DMS	Infrequently Flooded	X	
Shallow Marsh Shrub	SMS	Infrequently Flooded	X	
Wet Meadow Shrub	WMS	Infrequently Flooded	X	
Scrub-Shrub	SS	Infrequently Flooded	X	
Wooded Swamp	WS	Semipermanently Flooded	X	X
Floodplain Forest	FF	Seasonally Flooded	X	X
Populus Community	PC	Temporarily Flooded	X	X
Salix Community	SC	Infrequently Flooded	X	X
Lowland Forest	LF	Seasonally Flooded	X	X
Agriculture	AG	Seasonally Flooded		
Conifer	CN	Semipermanently Flooded	X	X
Plantation	PN	Seasonally Flooded	X	X
Upland Forest	UF	Temporarily Flooded	X	X
Developed	DV	Infrequently Flooded		
Grassland	GR	Infrequently Flooded	X	
Levee	LV	Infrequently Flooded	X	
Pasture	PS	Infrequently Flooded		
Roadside	RD	Infrequently Flooded	X	
Mudflat	MUD	Seasonally Flooded		
Sand Bar	SB	Temporarily Flooded		
Sand	SD	Infrequently Flooded		
No Photo Coverage	NPC	No Photo Coverage		

would be with the shrub and tree classes. These classes generally have a grassy understory. However, shrubs and trees are generally taller than grasses, so the density modifier would only apply to the percent canopy cover of the shrubs or trees in that polygon. Modifiers for density are as follows:

A = 0–33%   B = 34–66%   C = 67–90%   D = 91–100%

An example of an attribute in a polygon that delineates an area of Shallow Marsh Perennials (SMP) with a density of approximately 75% would be SMPC.

Height modifiers are only applicable to the tree classes. Shrub classes do not receive a height modifier because it is assumed that shrubs would have a height <20 ft. Modifiers for height are as follows:

1 = 0–20 ft. (young, regenerating stands)
2 = 21–50 ft. (maturing stands)
3 = >50 ft. (mature stands)

If height is used as a modifier, it will always follow that of density. An example of an attribute in a polygon that delineates *Populus* Community (PC) with a canopy cover of approximately 40% and an average height of approximately 70 ft. would be PCB3.

4. Automation

Once all the photointerpretation work is complete, the interpreted aerial photo overlays are converted to a digital format (automated) where they are geo-referenced to real world coordinates. During this process, the photointerpretation work is scanned into a computer where it becomes a raster image (.tif). The raster image is then referenced to the Earth's surface with ArcView Image Analysis (Environmental Systems Research Institute, Inc., Redlands, California) or Orthomapper (Image Processing Software, Madison, Wisconsin) software, which are geographic information system (GIS) programs. Once referenced, the raster image is then converted to vectors with another GIS software program known as ArcScan (fig.1–7).

After all the photointerpretation work is converted to a digital format, it is joined together and boundaries between adjacent polygons with the same attribute are removed (dissolved) and a single digital coverage is formed. The completed coverage is a land cover/land use map (fig. 1–8) that is converted into Arc/Info export files (.e00) or ArcView shapefiles (.shp). These files are then distributed through the Upper Midwest Environmental Sciences Center's Web site (*http://www.umesc.usgs.gov*) to resource managers, researchers, and analysts to help with natural resource management.

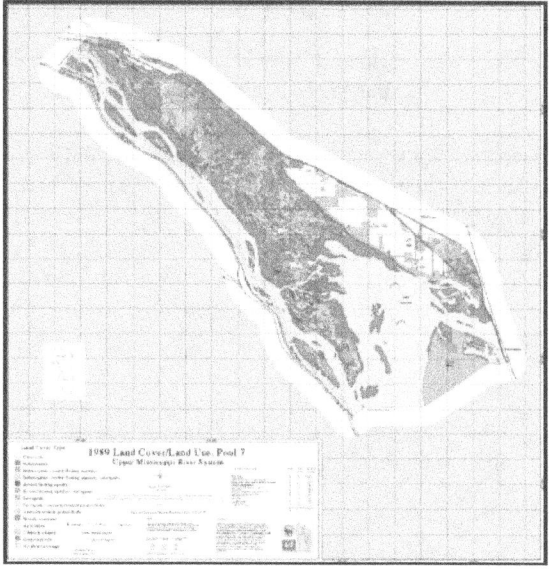

Figure 1–8. Example of a land cover/land use map created from photointerpreted color-infrared aerial photographs.

5. Accuracy Assessment

The accuracy of a land cover map can be assessed by identifying points on the map and then going into the field to determine if those particular locations have been properly classified. This process helps to correct systematic errors in

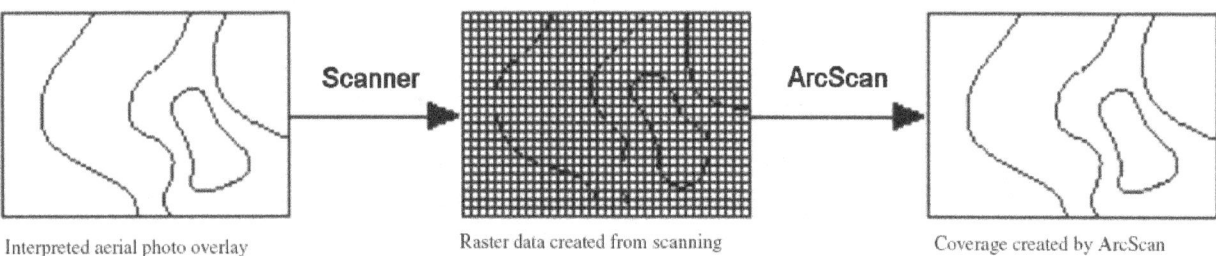

Interpreted aerial photo overlay   Raster data created from scanning   Coverage created by ArcScan

Figure 1–7. Example of photointerpretation work scanned or rasterized and then vectorized into a digital coverage with ArcScan.

the map. It also provides a quantitative measure of the overall accuracy of the map, as well as an indication of how well individual map classes are mapped.

A stratified random sampling design is typically used to conduct a statistically sound accuracy assessment and has become the standard approach for evaluating land cover maps (Environmental Systems Research Institute and others, 1994). In a stratified random sampling design, the land cover map is stratified by map class. Points are then randomly selected from each class. This method ensures that all classes are sampled and that each class is sampled in proportion to its occurrence on the landscape. After sampling points have been selected, a GPS unit is used to locate precise locations in the field. A description of the vegetation and other environmental features are recorded at each point.

The class determined in the field is then compared to the designation on the map for each point. If map and field determinations are conflicting, then an attempt is made to reconcile the difference. Differences may occur when points fall in transition zones between vegetation types or in areas that are too small to map. The GPS errors also account for some discrepancies. These kinds of errors, termed false errors, are corrected, reconciling the land cover map with the field determinations.

After false errors have been identified and reconciled, an error matrix is generated. This matrix, also called a misclassification or a contingency matrix, reports the frequency of agreement between the map classes labeled on the land cover map and the field determinations for each class. Based on the error matrix, the accuracy rate for classifying each map class can be determined, as well as the overall accuracy of the map.

Appendix 2.   General Wetland Vegetation Classification System

Appendix 2 is a detailed description of each of the 31 general classes in the General Wetland Vegetation Classification System. Each general class has at least one example of how that class may appear in the field, as well as 2- x 2-inch images extracted from the 1:24,000-scale color-infrared aerial photography. For each of the 2- x 2-inch images, there is a description of the map class signature (or photographic appearance). Also, an attribute is shown on each 2- x 2-inch image as it was noted by the photointerpreter. Each attribute contains the map code, followed by the density (A, B, C, or D) and height (1, 2, or 3) modifiers, when applicable (Table 2–1). For example, a class identified as Submersed Vegetation (SV) occurring at a density of 0–33% would be designated on the photograph by the code SVA.

Table 2-1.   The 31 general map classes together with their respective codes and possible density and height modifiers. Density and height modifiers designated by an X indicate that they apply to that map class.

Map class	Map code	Density[1]	Height[2]
Open Water	OW		
Submersed Vegetation	SV	X	
Rooted-Floating Aquatics	RFA	X	
Deep Marsh Annual	DMA	X	
Deep Marsh Perennial	DMP	X	
Shallow Marsh Annual	SMA	X	
Shallow Marsh Perennial	SMP	X	
Sedge Meadow	SM	X	
Wet Meadow	WM	X	
Deep Marsh Shrub	DMS	X	
Shallow Marsh Shrub	SMS	X	
Wet Meadow Shrub	WMS	X	
Scrub-Shrub	SS	X	
Wooded Swamp	WS	X	X
Floodplain Forest	FF	X	X
Populus Community	PC	X	X
Salix Community	SC	X	X
Lowland Forest	LF	X	X
Agriculture	AG		
Conifer	CN	X	X
Plantation	PN	X	X
Upland Forest	UF	X	X
Developed	DV		
Grassland	GR	X	
Levee	LV	X	
Pasture	PS		
Roadside	RD	X	
Mudflat	MUD		
Sand Bar	SB		
Sand	SD		
No Photo Coverage	NPC		

[1]Density modifiers are as follows: A = 0–33%, B = 34–66%, C = 67–90%, and D = 91–100%.

[2]Height modifiers are as follows: 1 = 0–20 ft. (young, regenerating stands), 2 = 21–50 ft. (maturing stands), 3 = >50 ft. (mature stands).

Open Water (OW)

Open Water (OW) represents the main channel and portions of lakes, ponds, and backwaters that remain permanently flooded all year and appear <10% vegetated. Areas that have >10% vegetation are classified into a general class that best represents that vegetation type, except in the instance of duckweed (*Lemna*, *Spirodela*, and *Wolffia*) and other nonrooted-floating aquatics. Because duckweed is free-floating, it can relocate day-to-day depending on current and wind direction. Therefore, any area of water containing dense duckweed will be classified as Open Water.

In Image A, the signature for water appears smooth and blue, but may range from light blue to black. Variation in color is typically because of water depth, turbidity, and sediment type. Generally, the clearer the water, the darker it appears. In an instance where duckweed covers the water (Image B), the signature appears white. It is unknown to the photointerpreter what lies beneath the duckweed, so it is attributed as Open Water. Image A was taken in August 2000, and Image B was taken in September 2000.

A
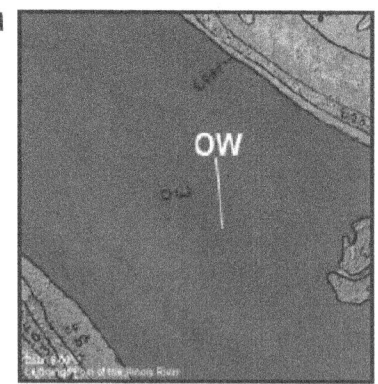

B
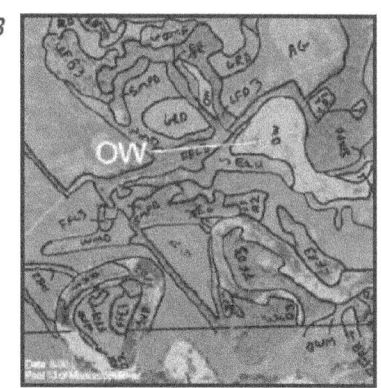

Submersed Vegetation (SV)

Submersed Vegetation (SV) represents portions of lakes, ponds, channel borders, or backwaters that appear >10% vegetated with vegetation growing and remaining underwater. This general class is dominated by submersed vegetation, but may have inclusions of nonrooted-floating aquatics, rooted-floating aquatics, or emergent vegetation. It generally grows between water depths of 0.5 and 2 m. This general class remains permanently flooded all year. Submersed vegetation that does not reach the water's surface may not be visible on the photographs and would be classified as OW.

The signature for submersed vegetation is generally dark gray/blue to black and appears discontinuous and clumped or gradational in the water. This can be seen in Image A. The signature in Image B also contains small white patches of duckweed. Here, the duckweed is positioned erratically enough to determine that submersed vegetation is present. Images A and B were taken in September 2000.

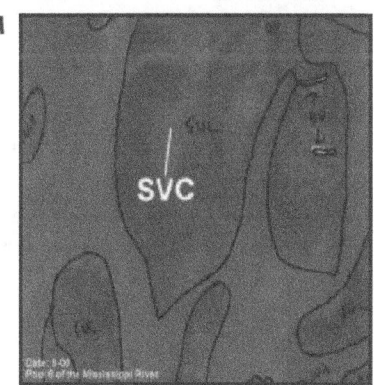

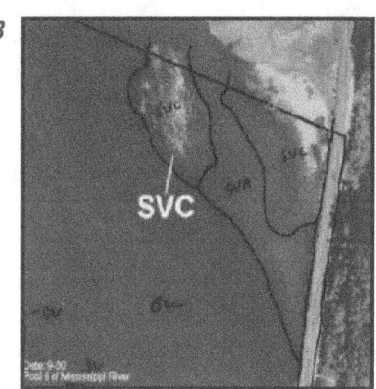

Rooted-Floating Aquatics (RFA)

Genus: *Nuphar*

Genus: *Nymphaea*

Rooted-Floating Aquatics (RFA) represent portions of lakes, ponds, marshes, backwaters, or channel borders that are >10% vegetated with water lilies (*Nymphaea* and *Nuphar*) or American Lotus (*Nelumbo*). This general class is dominated by rooted-floating aquatics, but may have inclusions of submersed, nonrooted-floating aquatics, or emergent vegetation. It is typically found growing between water depths of 0.25 and 2 m. This general class remains permanently flooded all year.

Images A and B show examples of the water lily signature. It lies on the water and appears flat, opaque, and pale pink. The signature in Image A contains small patches of dark blue water and white duckweed within the polygon. The signature in Image B appears solid pink with little duckweed present. Images A and B were taken in September 2000.

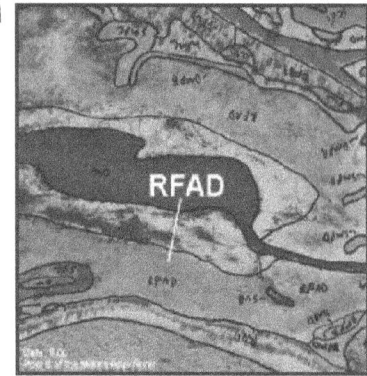

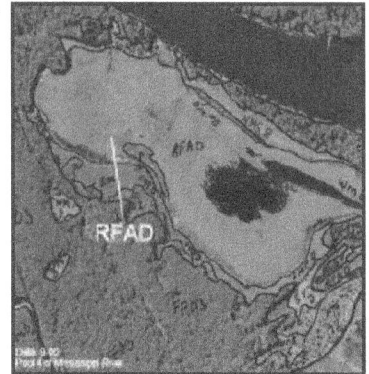

Genus: *Nelumbo*

Images C and D show examples of the American Lotus signature. It appears bright pink and rough. The signature in Image C contains small patches of dark blue water and white duckweed within the polygon. The signature in Image D appears a solid bright pink. Image C was taken in September 2000, and Image D was taken in August 2000.

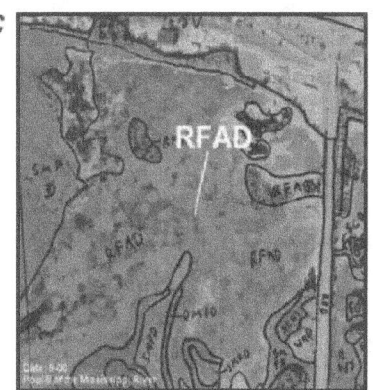

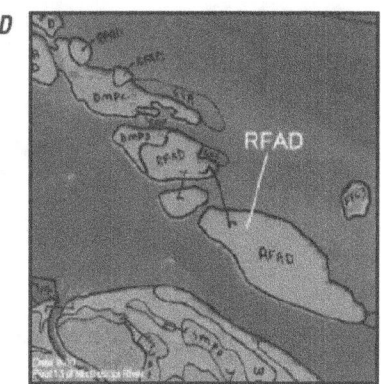

Deep Marsh Annual (DMA)

Genus: *Zizania*

Deep Marsh Annuals (DMA) represent portions of lakes, ponds, marshes, or backwaters that are >10% vegetated with wild rice (*Zizania*). This general class is dominated by wild rice, but may have inclusions of submersed, nonrooted-floating aquatics, rooted-floating aquatics, or emergent vegetation. It is typically found growing between water depths of 0.25 and 2 m with a silty or mucky bottom. This general class is semipermanently flooded throughout the year.

Images A and B show examples of the wild rice signature. It is generally light pink or peach and appears tall and fluffy. The darker blue areas visible within the wild rice signature are water. Images A and B were taken in August 2000.

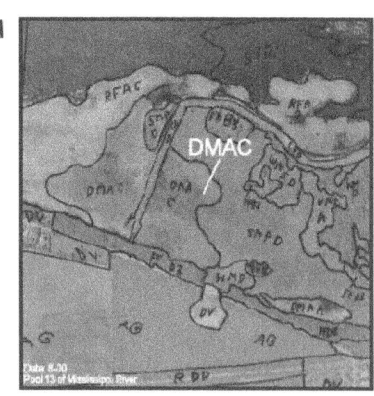

A

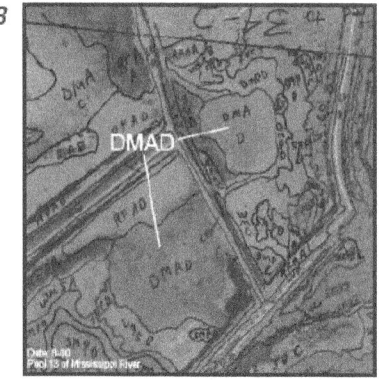

B

Deep Marsh Perennial (DMP)

Genus: *Pontederia*

Genus: *Sagittaria*

Deep Marsh Perennials (DMP) represent portions of lakes, ponds, marshes, or backwaters that are semipermanently flooded and >10% vegetated with persistent emergent vegetation dominated by pickerelweed (*Pontederia*), arrowhead (*Sagittaria*), cattail (*Typha*), or bur-reed (*Sparganium*). This general class may have inclusions of submersed, nonrooted-floating aquatics, rooted-floating aquatics, or other emergent vegetation and is typically found growing in water up to 1 m deep.

Images A and B show examples of the arrowhead signature. Arrowhead generally grows at the water's edge and appears as pink to red velvety clumps. Pickerelweed is similar in signature to that of arrowhead, but generally appears deeper red. The arrowhead signature in Images A and B contain patches of white duckweed. Images A and B were taken in September 2000.

A

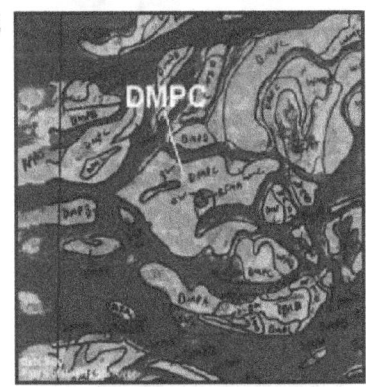

B

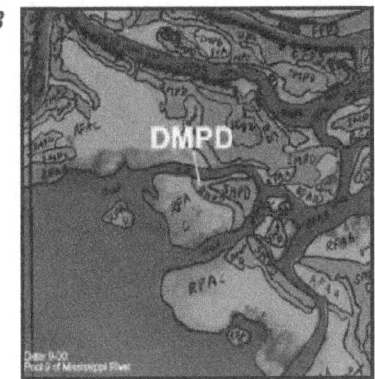

Genus: *Typha*

Genus: *Sparganium*

Images C and D show examples of the cattail and bur-reed signatures. The cattail and bur-reed signatures are similar. They both appear textured and deep red to brown. The most prominent distinguishing characteristic of the two signatures is that cattail often grows clonally, whereas bur-reed grows irregularly and near the water's edge. Field reconnaissance is often needed to accurately differentiate between the two signatures. Images C and D were taken in September 2000.

C

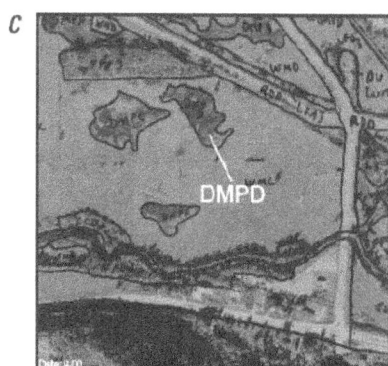

D

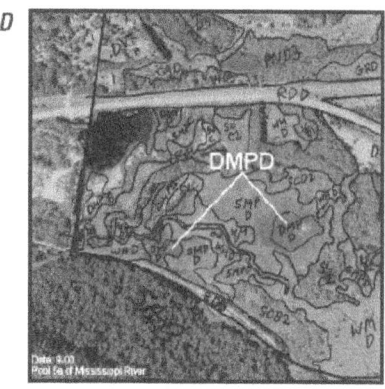

Shallow Marsh Annual (SMA)

Genus: *Bidens*

Genus: *Cyperus*

Shallow Marsh Annuals (SMA) represent portions of lakes, ponds, backwaters, mudflats, or shorelines that are seasonally flooded and >10% vegetated with annual (non-persistent) emergent vegetation. Common vegetation types include wild millet (*Echinochloa*), pinkweed (*Polygonum*), spike-rush (*Eleocharis*), red-root flatsedge (*Cyperus*), and beggarticks (*Bidens*). This general class may have inclusions of submersed, nonrooted-floating aquatics, or persistent emergent vegetation. It is typically found growing on soils that are saturated or inundated by water up to 0.2 m deep.

Images A and B show examples of the shallow marsh annual signature. With the exception of wild millet, the signature most often appears, short, fluffy, and pale pink. Image A contains patches of blue water within the polygon. Image B contains areas of gray mud throughout the polygon. Images A and B were taken in August 2000.

A
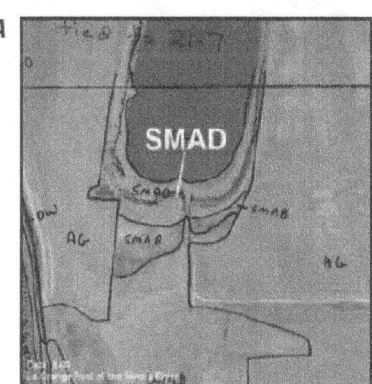

B
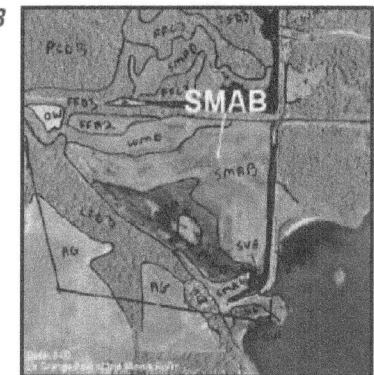

Genus: *Echinochloa*

Genus: *Eleocharis*

Image C shows the wild millet signature. It is generally found growing near the water's edge and appears tall
and bright red. The signature in Image C also contains small
patches of dark blue water, as well as, other light pink shallow
marsh annuals. Wild millet is a shallow marsh annual, therefore both signatures become part of the same polygon. Image
C was taken in September 2000.

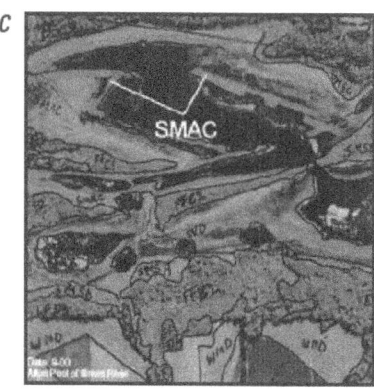

Shallow Marsh Perennial (SMP)

Genus: *Lythrum*

Genus: *Phragmites*

Shallow Marsh Perennials (SMP) represent portions of lakes, ponds, backwaters, or shorelines that are seasonally flooded and >10% vegetated with persistent emergent vegetation. The SMP denote the transition zone between deep marsh perennials and wet meadow. Common vegetation types include bulrush (*Scirpus*), purple loosestrife (*Lythrum*), giant reed grass (*Phragmites*), and smartweed (*Polygonum*). This general class may have inclusions of submersed, nonrooted-floating aquatics, or other emergent vegetation. It is typically found growing on soils that are saturated or inundated by water up to 0.2 m deep.

Images A–D show examples of the shallow marsh perennial signature. As seen in these images, a great deal of variation occurs within the shallow marsh perennial signature depending upon the vegetation type. It may range from grayish-green to orange or red and generally appears thick and textured.

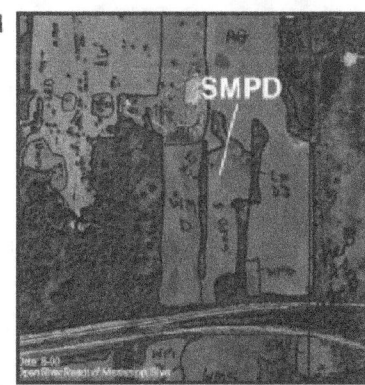

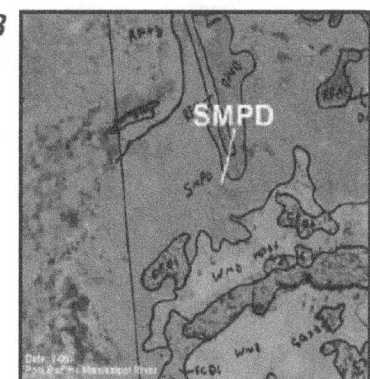

Genus: *Scirpus*

Genus: *Scirpus*

In Image A, the signature is thick and appears orange and pink. This signature represents purple loosestrife and smartweed vegetation types. Purple loosestrife generally appears orange, whereas smartweed ranges from light to bright pink.

In Images B and C, the signature is also thick but appears gray and pink. These signatures represent the bulrush and giant reed grass vegetation types. Bulrush can range from grayish-green to red, whereas giant reed grass generally appears gray.

In Image D, the signature again appears thick and red. Image D also contains a few patches of white duckweed within the polygon. This signature is primarily comprised of bulrush. Image A was taken in August 2000, and Images B-D were taken in September 2000.

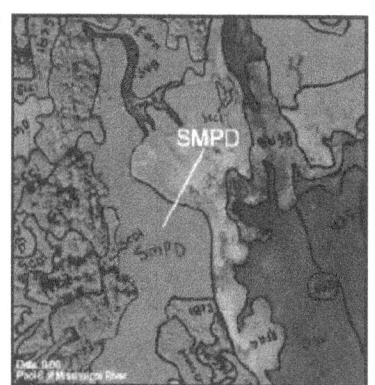

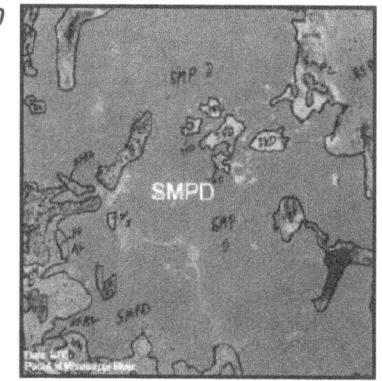

Sedge Meadow (SM)

Genus: *Carex*

Sedge Meadow (SM) represents areas around lakes, ponds, backwaters, and along shorelines that are temporarily flooded and >10% vegetated with sedges. Sedge meadows are generally dominated by *Carex*. This general class may have inclusions of moist soil grasses and forbs or persistent emergent vegetation. It is typically found growing on saturated soils comprised of peat or muck, but will often grade into shallow marshes or wet meadows.

Image A shows an example of the sedge meadow signature. It generally appears smooth and pink. The polygon drawn for this general class is small. This general class is rarely used with photographs taken at 1:24,000. At this scale, areas dominated by sedges generally blend in with the wet meadow class. Image A was taken in September 2000.

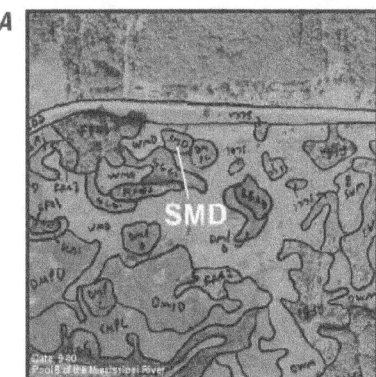

A

Wet Meadow (WM)

Genus: *Leersia*

Genus: *Phalaris*

Wet Meadow (WM) represents lowland areas that are >10% vegetated with perennial grasses and forbs. Common vegetation types include reed canary grass (*Phalaris*), rice cutgrass (*Leersia*), and goldenrod (*Solidago*). This general class may have small inclusions of woody vegetation, sedges, or emergent vegetation, such as smartweed or purple loosestrife. It is typically found growing on saturated soils and is often considered the transition zone between aquatic communities and uplands.

The signature for wet meadow can vary depending on the vegetation type or types. Image A shows a monotypic stand of rice cutgrass. This signature is bright pink and smooth. Image B shows a monotypic stand of reed canary grass. It appears medium pink with white speckles. Both images are in the transition zone, between shallow marsh perennials and drier wooded areas. Images A and B were taken in September 2000.

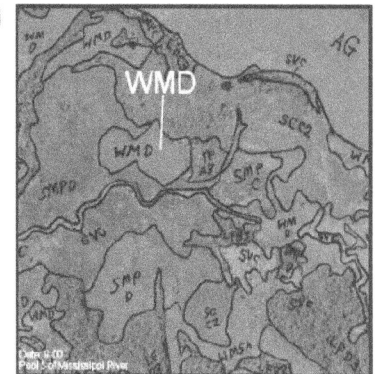

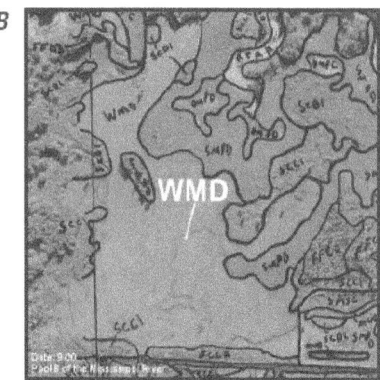

Stands that are not monotypic can appear in a variety of colors, including gray, brown, pink, and red. Their range in color is generally because of the type of vegetation present, as well as, how saturated the soil may be. Image C shows an example of a signature comprised of a mix of grasses and forbs. The signature appears rough with several shades of pink. This area of wet meadow is in an open area of floodplain forest. Image C was taken in August 2000.

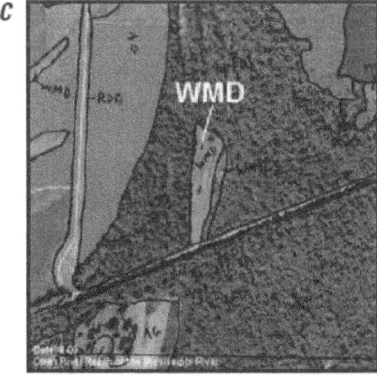

Deep Marsh Shrub (DMS)

Deep Marsh Shrubs (DMS) represent areas in or around lakes, ponds, backwaters, or shorelines that are >25% vegetated with semipermanently flooded shrubby vegetation. Common vegetation types include buttonbush (*Cephalanthus*) and water willow (*Decodon*). This general class may have inclusions of submersed, nonrooted-floating aquatics, rooted-floating aquatics, or emergent vegetation. It is typically found growing in shallow water.

Images A and B show examples of the deep marsh shrub signature. They generally appear red and speckled or beady. The signature in Image A is deeper red than that in Image B. This is primarily because the shrubs in Image A are denser. Also, the deep marsh shrubs in Image A may be a little larger than those in Image B. In both images, dark blue water is interspersed between the shrubs. Images A and B were taken in August 2000.

A

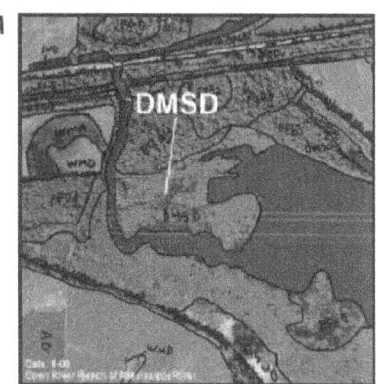

B

Shallow Marsh Shrub (SMS)

Shallow Marsh Shrubs (SMS) represent areas near the shoreline or around lakes, ponds, and backwaters that are >25% vegetated with seasonally flooded shrubby vegetation. It typically grows with mixed emergents, grasses, and forbs. This general class tends to be drier than deep marsh shrubs, but wetter than wet meadow shrubs. Sandbar willow (*Salix*) may be growing in this mix of shrubby vegetation. Shallow marsh shrubs are typically found growing on soils that are saturated or inundated with little water.

Images A and B show examples of the SMS signature. The SMS are generally sporadic and appear pink to red and speckled or beady. In Image A, the shallow marsh shrubs are small and barely visible. In Image B, the shrubs can be clearly seen. Throughout both polygons, the pink groundlayer consists of a mix of grasses, forbs, and emergent vegetation. Images A and B were taken in August 2000.

A

B
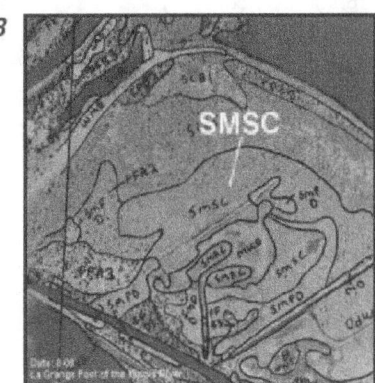

Wet Meadow Shrub (WMS)

Wet Meadow Shrubs (WMS) represent lowland areas that are >25% vegetated with temporarily flooded shrubby vegetation. This general class tends to be drier than shallow marsh shrubs, but wetter than scrub-shrubs, and typically grows with a mix of sedges, grasses, and forbs. Common vegetation types include alder (*Alnus*), elder (*Sambucus*), false indigo (*Amorpha*), dogwood (*Cornus*), and willow (*Salix*). Wet meadow shrubs are typically found growing on saturated soils.

Images A and B show examples of the WMS signature. The WMS are generally sporadic and appear pink to red and speckled or beady. In Images A and B, the darker pink areas represent the shrubs and the lighter pink areas represent the groundlayer consisting of a mix of grasses and forbs. Images A and B were taken in August 2000.

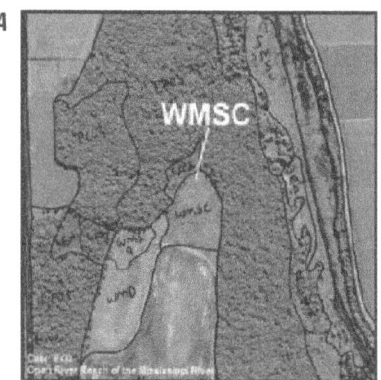

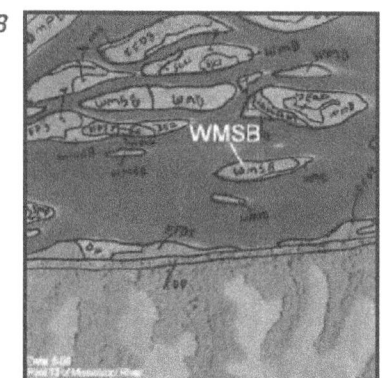

Scrub-Shrub (SS)

Scrub-Shrubs (SS) represent upland areas that are >25% vegetated with infrequently flooded shrubby vegetation. This general class is the driest of the shrub classes and typically grows with a mix of grasses and forbs on drier soils.

Images A and B show examples of the scrub-shrub signature. Scrub-shrubs are generally sporadic and appear pink to red and speckled or beady. In Images A and B, the scrub-shrubs are growing on a hill where soils are drier. In Image A, the shrubs appear pink with a grassy groundlayer that is gray. In Image B, the shrubs appear dark pink with a grass and forb groundlayer that is light pink. Image A was taken in September 2000, and Image B was taken in August 2000.

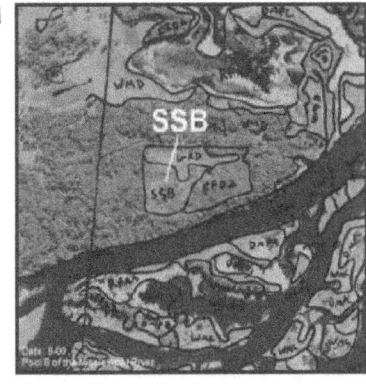

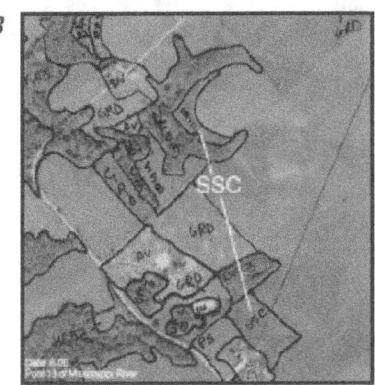

Wooded Swamp (WS)

Wooded Swamp (WS) represents areas in or around shallow lakes, ponds, oxbows, or backwaters that are >10% vegetated with semipermanently flooded forests. Common vegetation types include bald cypress (*Taxodium*), water tupelo (*Nyssa*), sourgum (*Nyssa*), and black ash (*Fraxinus*). This general class is most common in southern reaches of the Upper Mississippi River System. It may have inclusions of submersed, nonrooted-floating aquatics, rooted-floating aquatics, or emergent vegetation. It is typically found growing in shallow water.

Image A shows an example of a wooded swamp signature. It generally ranges from red or pink to brown. The signature in Image A shows the trees pink with dark blue patches of water interspersed between them. Image A was taken in August 2000.

A

Floodplain Forest (FF)

Floodplain Forest (FF) represents areas on islands, near the shoreline, or around lakes, ponds, and backwaters that are >10% vegetated with seasonally flooded forests. These forests are predominantly silver maple (*Acer*), but also include elm (*Ulmus*), cottonwood (*Populus*), black willow (*Salix*), and river birch (*Betula*). This general class is typically found growing at or near the water table where it becomes inundated from spring flooding and high-water events.

Images A and B show examples of a floodplain forest signature. In Image A, the trees appear red with dark blue water surrounding them. In Image B, the trees also appear red; however, they are not as prominent. This is primarily because the trees in Image B are smaller than the trees in Image A. The floodplain forest in Image B is also adjacent to shallow marsh perennials. Image A was taken in September 2000, and Image B was taken in August 2000.

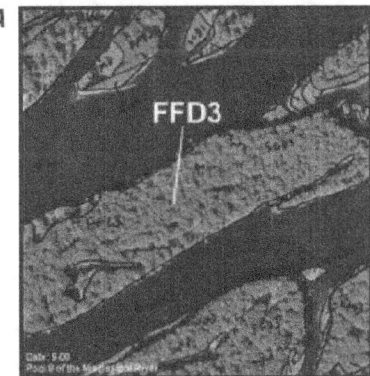

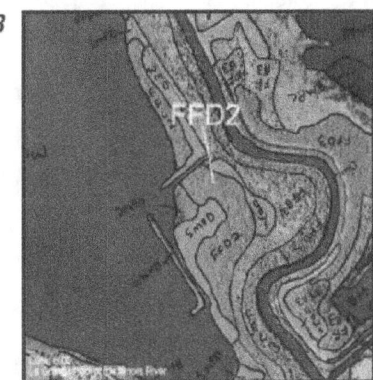

Populus Community (PC)

Populus Community (PC) represents lowland areas that are >10% vegetated with seasonally flooded cottonwood trees. These forests are >50% cottonwood (*Populus*) and may include other floodplain and lowland forest types. This general class is typically a pioneering species of disturbed areas and is generally found growing on moist soils. *Populus* communities are tall and often grow monotypically, as well as adjacent to or along with floodplain forest or lowland forest types.

Images A and B show examples of the PC signature. The signature generally ranges from light gray to purple. In both Images A and B, the stands are monotypic and are adjacent to floodplain forest types. In Image A, the cottonwood trees are purple. In Image B, the cottonwood trees are light gray. Image A was taken in August 2000, and Image B was taken in September 2000.

A

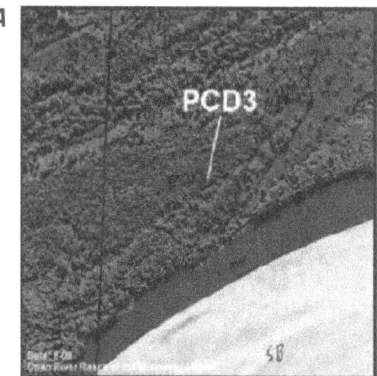

B

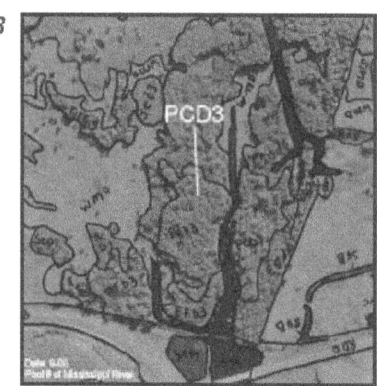

Salix Community (SC)

Salix Community (SC) represents areas near the shoreline or around lakes, ponds, and backwaters that are >10% vegetated with seasonally flooded willow trees or shrubs. These forests or shrub communities are >50% willow (*Salix*) and may include other floodplain forest types. This general class typically grows with an emergent, grass, and/or forb understory on moist and saturated soils.

Images A and B show examples of the SC signature, which is often dense, textured, and light to medium pink. In Image A, the willow shrubs are medium pink. The lighter pink areas represent the grass/forb understory. Floodplain forest types surround the willow shrubs in Image A. In Image B, the willow shrubs are also medium pink. The lighter pink patches in the polygon represent a mix of grasses, forbs, and emergents. Image A was taken in September 2000, and Image B was taken in August 2000.

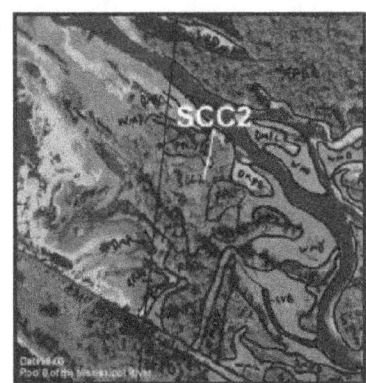

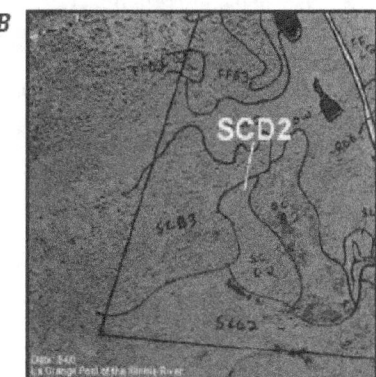

Lowland Forest (LF)

Lowland Forest (LF) represents areas along the river-banks and within the floodplain that are drier than floodplain forest sites and are >10% vegetated with temporarily flooded forests. Common vegetation types include pecan (*Carya*), hickory (*Carya*), river birch (*Betula*), sycamore (*Platanus*), and red/black oak (*Quercus*). This general class is most common in southern reaches of the Upper Mississippi and Illinois River Systems and is typically found growing on moist, well-drained soils.

Images A and B show examples of the lowland forest signature. In both images, the lowland forest appears red to dark red. The small pink patches throughout the lowland forest in Image A and the lower portion of Image B represent the grass/forb understory. Both stands of lowland forest exist within the floodplain and are surrounded by agriculture. Images A and B were taken in August 2000.

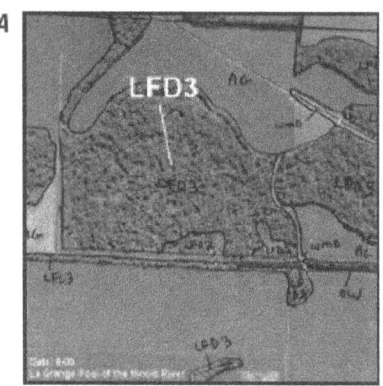

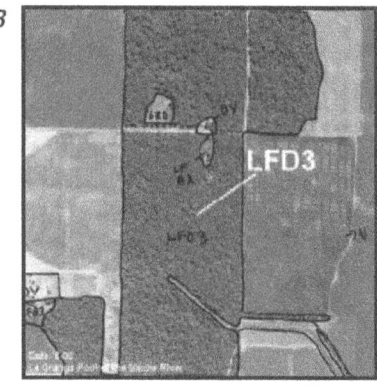

Agriculture (AG)

Agriculture (AG) represents all obviously cultivated fields for crops. This general class may include transitional fallow fields that show evidence of tilling. Because of a large floodplain, vast agricultural areas are common in the southern reaches of the Upper Mississippi and Illinois River Systems. Agriculture is generally considered infrequently flooded, however, it is not uncommon to find cultivated fields within seasonally or temporarily flooded areas.

Images A and B show examples of the agriculture signature. The signature is generally uniform, smooth, and ranges from white to red. In Image A, the signature appears pink and white, with areas of harvest in white and areas of standing crop in pink. This example is located within an area that typically floods each year. Image B shows an agricultural area surrounded by a small town. Here the signature is uniform, smooth, and pink. Image A was taken in August 2000, and Image B was taken in September 2000.

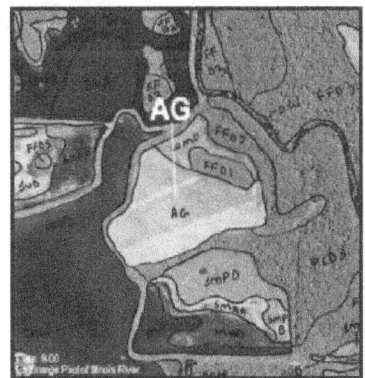

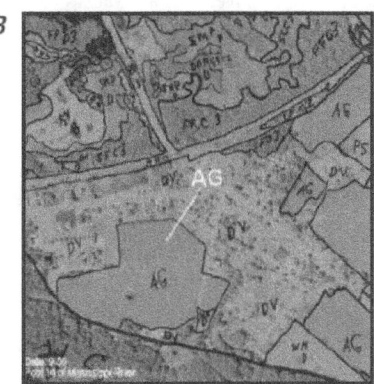

Conifer (CN)

Conifers (CN) represent forested areas that are >10% vegetated with natural or seminatural evergreen communities. These communities are typically pine, but may also include cedar. This general class is infrequently flooded and is typically found growing in lowland or upland situations where the soils are well drained.

Images A and B show examples of the conifer signature. In general, the signature appears dark red to brown. In Image A, the conifers appear brown. The red patches within the signature represent deciduous trees. The conifers in Image A are located on a hill and are surrounded by areas of grassland and agriculture. In Image B, the conifers are located on a slope between the river and a small-developed area. Here the signature appears dark red. Image A was taken in August 2000, and Image B was taken in September 2000.

A

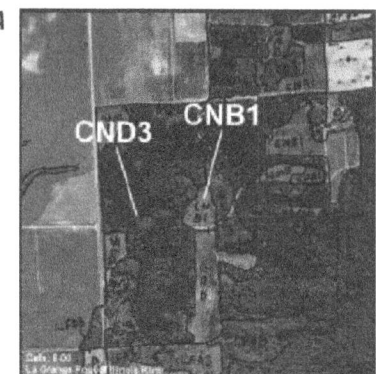

B

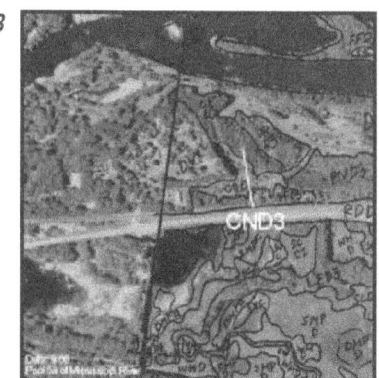

Plantation (PN)

Plantation (PN) represents forested areas that are >10% vegetated with commercially grown evergreen plantations, large nurseries, or orchards. This general class typically consists of red or white pine (*Pinus*), but may include other coniferous or deciduous trees. Plantations are infrequently flooded and are typically found growing in lowland or upland situations where the soils are well drained.

Images A and B show examples of the plantation signature. The signature visibly shows the trees growing in rows. In general, the coniferous plantations appear dark red to brown, whereas deciduous plantations appear red. In both Images A and B, the trees are coniferous. They appear in rows and are brown. Both plantations are in areas surrounded by grasslands and agriculture. Images A and B were taken in August 2000.

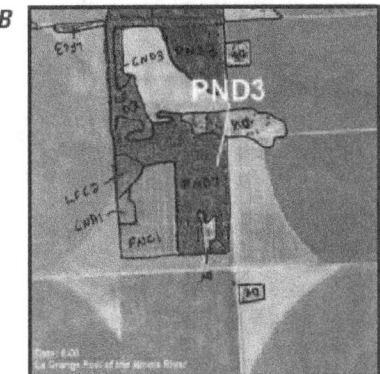

Upland Forest (UF)

Upland Forest (UF) represents forested areas that are >10% vegetated with forests growing on hills near the edge of the floodplain, or out of the floodplain. This general class typically consists of red or white oak (*Quercus*), hickory (*Carya*), elm (*Ulmus*), and other deciduous trees. Upland forests are infrequently flooded and are typically found growing in upland situations where soils are dry.

Images A and B show examples of the upland forest signature. In both images, the upland forest appears red to dark red. The upland forest in Image A is located between an agricultural area and a developed area. The small gray patches within the upland forest polygon represent grasses. Image B is located on a large hill and lies adjacent to agricultural fields. Images A and B were taken in August 2000.

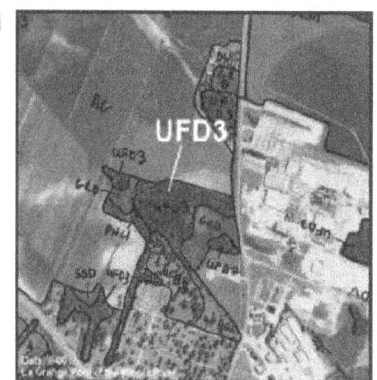

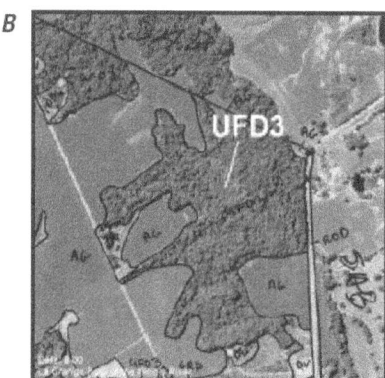

Developed (DV)

Developed (DV) represents areas that are predominantly artificial in nature. This general class includes residential homes in populated areas, homesteads in rural settings, farm-steads, industrial complexes, parks, locks and dams, marinas, boat launches, rip-rap, and newly constructed artificial islands. Most developed areas are considered infrequently flooded, however, rip-rap and newly constructed artificial islands may be seasonally or temporarily flooded.

Images A and B show examples of two different types of the developed area signature. Image A shows an example of an industrial complex. Note all of the large industrial buildings. Image B is an example of a farmstead. The polygon was delin-eated only around the buildings. Agricultural fields surround the farmstead. Image A was taken in September 2000, and Image B was taken in August 2000.

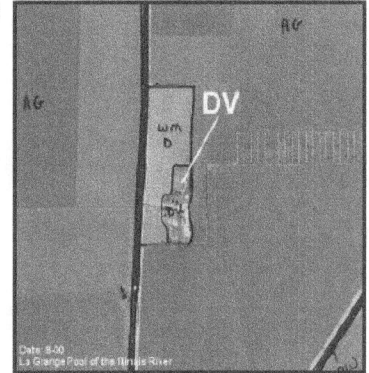

Image C shows the signature for a newly constructed artificial island. It appears white and smooth. Image D shows the signature for rip-rap. It also appears white, but may appear smooth or rough in texture. Once rip-rap and newly constructed artificial islands become >10% vegetated, they will transition into another class, such as WM. Images C and D were taken in September 2000.

C

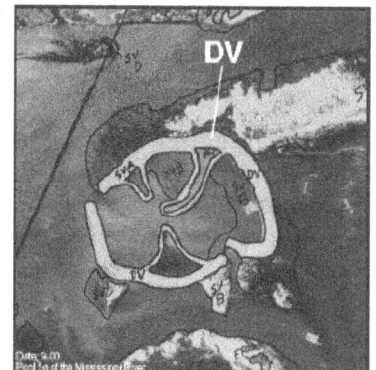

D

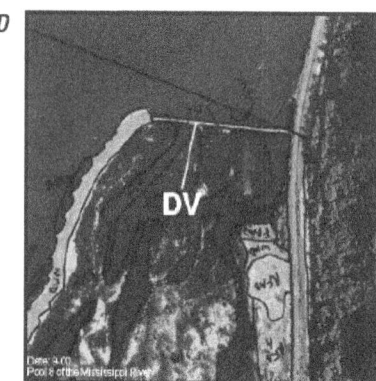

Grassland (GR)

Grassland (GR) represents drier upland areas that are >10% vegetated with perennial grasses and forbs. This general class may include fallow fields, sand prairies, and shrubby vegetation <25%. It generally exists near other upland types, such as scrub-shrubs or upland forest. Grasslands are infrequently flooded and are typically found growing where soils are dry.

Images A and B show examples of the grassland signature. It generally ranges from grayish-green to pink. The grassland in Image A is an example of a sand prairie. It appears gray and is surrounded by other shrub and forest types. Image B is located on the slope of a hill where GR appears grayish-green. It is near agricultural fields and a small, developed area. Image A was taken in September 2000, and Image B was taken in August 2000.

A

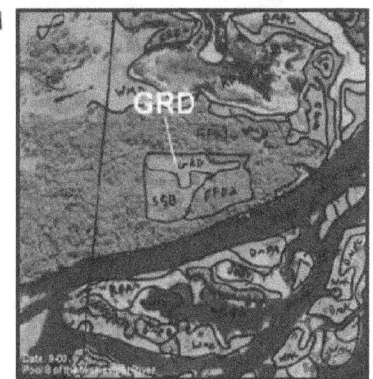

B

Levee (LV)

Levee (LV) represents all continuous dikes or embankments designed for flood protection. This general class is elevated and is typically covered with a mix of perennial grasses and forbs. Occasionally, shrubs may grow along or atop these structures. Levees are more commonly found in the southern reaches of the Upper Mississippi River System and are considered infrequently flooded.

Images A and B show examples of the levee signature. It generally ranges from grayish-green to pink. In both Images A and B, the levee appears pink. The white line running through the middle of the structure is a gravel path atop the levee. Commonly, levees are constructed to prevent water from entering agricultural land. This can be seen in Image B. Images A and B were taken in August 2000.

A
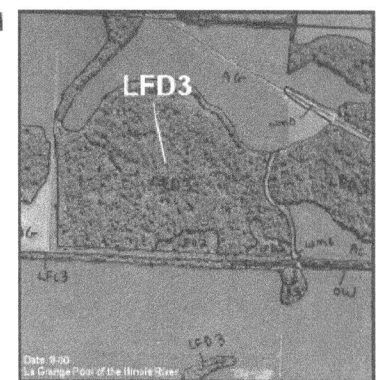

B
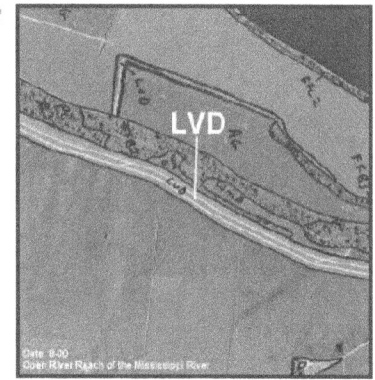

Pasture (PS)

Pasture (PS) represents areas used for the production of livestock. This general class typically grows with a mix of perennial grasses and forbs used for pasturing. Grasses and forbs are generally grazed and are maintained relatively short. Some of these grasses and forbs may also be hayed. Scattered shrubs (<25%) and trees (<10%) may be present. Pastures are considered infrequently flooded.

Images A and B show examples of the pasture signature. It generally ranges from grayish-green to pink and may appear mottled because of animal use. A fence line can often be seen surrounding the pastured area. In Image A, the pasture is located adjacent to a farmstead and appears gray. In Image B, the pastures appear pink with white mottling. These pastures are just outside an urban area and are surrounded by agricultural fields. Image A was taken in August 2000, and Image B was taken in September 2000.

A

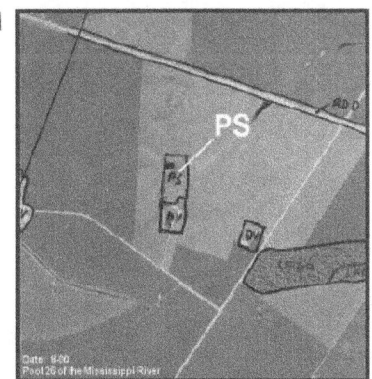

B

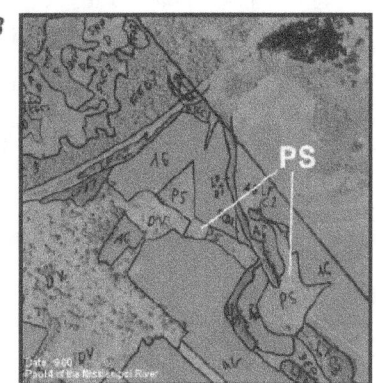

Roadside (RD)

Roadside (RD) represents roads, highways, and railroads along with their respective rights-of-way. These rights-of-way are typically covered with a mix of perennial grasses, forbs, and shrubs (< 25%). Scattered trees (<10%) may also be present. Typically, RD is used to classify only major, rural roadways, leaving out small narrow roads and trails. Roads within developed areas are mapped as part of the DV general class. Roadside is considered infrequently flooded.

Images A and B show examples of the roadside signature. The rights-of-way generally range from grayish-green to pink and are adjacent to a road or railway. Image A is an example of a railway with the rights-of-way appearing pink and gray. Image B is an example of a highway with the roadside and median appearing pink. Image A was taken in September 2000, and Image B was taken in August 2000.

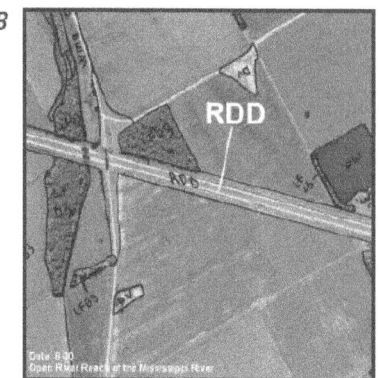

Mudflat (MUD)

Mudflats (MUD) represent portions of lakes, ponds, backwaters, or shorelines that are seasonally flooded and exposed with nonvegetated mud. This general class may have small inclusions (<10%) of persistent or nonpersistent emergent vegetation, sedges, grasses, or forbs. If exposed long enough, mudflats that remain moist will usually transition into the SMA class.

Images A and B show examples of the mudflat signature. It generally appears light to dark gray and smooth, but may contain a ripple effect. Both Images A and B are examples of mudflats that appear smooth and light gray. Typically, mudflats are mapped when the water recedes and nonpersistent emergent vegetation has not yet grown. Images A and B were taken in August 2000.

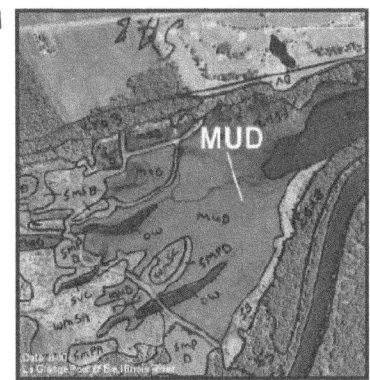

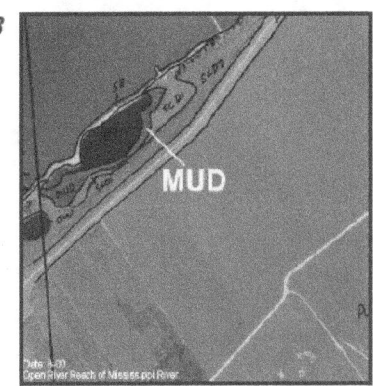

Sand Bar (SB)

Sand Bar (SB) represents areas that are temporarily flooded and exposed with nonvegetated sand flats. They are typically found in or near the main channel and are often associated with wing dams, shorelines, and islands. This general class may have small inclusions of grasses or forbs (<10%) or shrubs (<25%), but usually does not support plant life.

Images A and B show examples of the sand bar signature. It generally appears white, however, when wetter it may appear light gray. In Image A, the sand bar is in the main channel and appears both white and light gray. In Image B, the sand bar is downstream of an island and appears white. Images A and B were taken in August 2000.

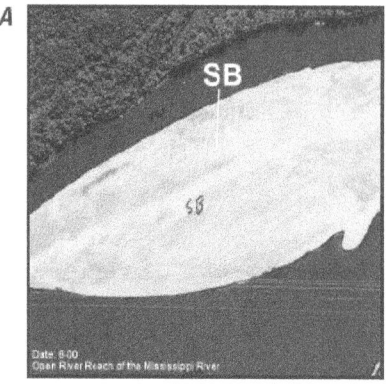

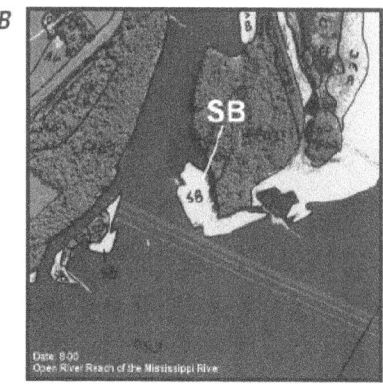

Sand (SD)

Sand (SD) represents areas that are infrequently flooded with nonvegetated sand. It typically includes sand spoil banks, beaches, and other sandy areas that are upland. This general class may have small inclusions of grasses or forbs (<10%), trees (<10%), or shrubs (<25%).

Images A and B show examples of the sand signature. It generally appears white and contains height. Images A and B are examples of sand spoil banks. Both contain height and appear white. The sand spoil bank in Image A also includes a few scattered trees that appear red. Images A and B were taken in September 2000.

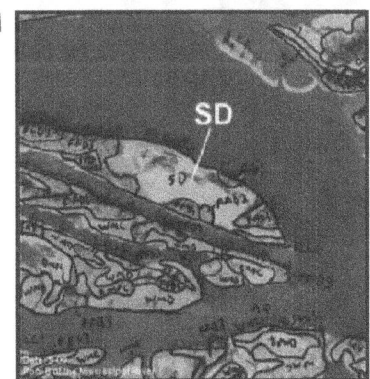

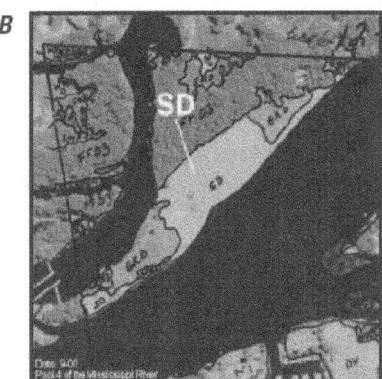

No Photo Coverage (NPC)

No Photo Coverage (NPC) represents gaps in the photography because of an incomplete data set, or areas that are obscured by clouds or shadows.

Image A shows an example of an area obscured by cloud cover. The clouds appear white with black shadows from the clouds on the ground. Image B shows a gap in the photography coverage. This is also known as a holiday. The photography was to cover the area connecting the highway (*top, left*) with the lock and dam (*top, right*). It can be seen that coverage of this area is not complete. The black strip on top of the image represents the photo's edge. Images A and B were taken in September 2000.

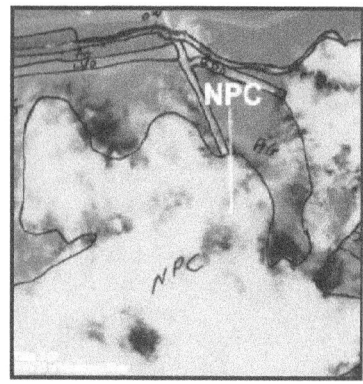

Appendix 3.   Classification Key for the General Wetland Vegetation Classification System

Below is a dichotomous key for the General Wetland Vegetation Classification System. Description of the codes are found in Table 1–1 of Appendix 1.

1a Vegetation <10% of the area   2

    2a Aquatic—Open water, or Lemnaceae sparse enough to see <10% submerged vegetation present, or Lemnaceae too dense to see submerged vegetation   OW

    2b Terrestrial   3

        3a Residential homes, homesteads in rural settings, farmsteads, industrial complexes, parks, locks and dams, marinas, boat launches, rip-rap, or newly constructed artificial islands   DV

        3b Exposed mud or sand   4

            4a Mudflat   MUD

            4b Sand   5

                5a Sand bar   SB

                5b Sand dunes, sand spoil banks, beaches, and other sandy areas that are upland   SD

1b Vegetation >10% of the area (not including Lemnaceae)   6

    6a Includes residential homes, homesteads in rural settings, farmsteads, or parks   DV

    6b Does not include residential homes, homesteads in rural settings, farmsteads, or parks   7

        7a Shrub cover <25% of the area and tree cover <10% of the area   8

            8a Submerged vegetation >10% of the vegetation; all other life forms <10% of the vegetation   SV

            8b At least one nonsubmerged species >10% of the vegetation, submerged vegetation may be present or absent   9

                9a Rooted-floating aquatics (e.g., *Nelumbo, Nymphaea, Nuphar*) >50% of the vegetation   RFA

                9b Annual or perennial emergents or perennial grasses or forbs >50% of the vegetation   10

                    10a Annual or perennial emergents >50% of the vegetation   11

                        11a Rooted-floating aquatics >10% of the vegetation   DMP

                        11b Rooted-floating aquatics <10% of the vegetation   12

                            12a Deep marsh species (e.g., *Pontederia, Sagittaria, Sparganium, Typha, Zizania*) >50% of the vegetation   13

                                13a Annuals (e.g., *Zizania*)   DMA

                                13b Perennials (e.g., *Pontederia, Sagittaria, Sparganium, Typha*)   14

                                    14a One or two species; may include rooted-floating aquatics >10% of the vegetation   DMP

                                    14b One species >50% of the vegetation and species other than rooted-floating or deep marsh >10% of the vegetation; or three or more deep marsh species   SMP

                            12b *Carex* or shallow marsh species (e.g., *Bidens, Cyperus, Echinochloa, Eleocharis, Lythrum, Phragmites, Scirpus*) >50% of the vegetation   15

                              15a *Carex* >50% of the vegetation   SM

                              15b Shallow marsh species >50% of the vegetation   16

                                16a Annuals (e.g., *Bidens, Cyperus, Echinochloa, Eleocharis*)   SMA

                                16b Perennials (e.g., *Lythrum, Phragmites, Scirpus*)   17

                                  17a *Lythrum* >50% of the vegetation   18

                                    18a Only *Lythrum* present   SMP

                                    18b *Lythrum* >50% of the vegetation and one or more species >10% of the vegetation   WM

                              17b Shallow marsh species other than *Lythrum* >50% of the vegetation   19

19a One species or a combination of species >50% of the vegetation; except when *Phragmites* >50% of the vegetation and *Phalaris* >10% of the vegetation ..... SMP

19b *Phragmites* >50% of the vegetation and *Phalaris* >10% of the vegetation ..... WM

10b Perennial grasses or forbs >50% of the vegetation 20

20a Landscape altered for human use 21

21a Areas for agricultural or livestock use 22

22a Cultivated fields for crops ..... AG

22b Pastured area used for production of livestock ..... PS

21b Areas not for agricultural or livestock use 23

23a Roads or railroads including grasses, forbs, or shrubs in rights-of-way ..... RD

23b Levees (continuous dikes or embankments) LV

20b Landscape not altered for human use 24

24a Wet soils (e.g., *Amaranthus, Leersia, Phalaris, Solidago, Spartina*) ..... WM

24b Dry soils ..... GR

7b Shrub cover >25% of the area or tree cover >10% of the area 25

25a Shrub cover >25% of the area and tree cover <10% of the area 26

26a *Salix* >50% of the vegetation ..... SC

26b Other shrubs >50% of the vegetation 27

27a Shrubs growing in standing water or with annual or perennial emergents 28

28a Shrubs (e.g., *Cephalanthus, Decodon*) growing in standing water or with deep marsh species (e.g., *Pontederia, Sagittaria, Sparganium, Typha, Zizania*) ..... DMS

28b Shrubs growing with shallow marsh species (e.g., *Bidens, Cyperus, Echinochloa, Eleocharis, Lythrum, Phragmites, Scirpus*) ..... SMS

27b Shrubs growing with perennial grasses or forbs 29

29a Wet soils (e.g., *Alnus, Cornus, Sambucus*) ..... WMS

29b Dry soils ..... SS

25b Tree cover >10% of the area 30

30a Cultivated areas (e.g., orchards or pine plantations) ..... PN

30b Noncultivated areas 31

31a *Populus* or *Salix* >50% of the vegetation 32

32a *Populus* >50% of the vegetation ..... PC

32b *Salix* >50% of the vegetation ..... SC

31b Other trees >50% of the vegetation 33

33a Coniferous trees >50% of the vegetation (e.g., *Pinus, Juniperus*) ..... CN

33b Deciduous trees >50% of the vegetation 34

34a Trees growing in standing water (e.g., *Taxodium, Nyssa*) ..... WS

34b Trees not growing in standing water 35

35a Trees growing on wet soils 36

36a Trees growing on alluvial soils; usually dominated by *Acer* ..... FF

36b Trees growing on moist, well-drained soils; usually dominated by *Quercus* ..... LF

35b Trees growing on dry soil ..... UF

Appendix 4.   Predominant Species and Common 31 General Classes Associated with the Genera Referred to in this Handbook

Genus	Species represented	Common classes
Acer	A. negundo, A. rubrum, A. saccharinum	FF, LF, UF
Alnus	A. glutinosa, A. serrulata	WMS
Amaranthus	A. albus, A. rudis, A. tuberculatus	WM
Amorpha	A. fruiticosa	WMS
Betula	B. nigra	FF, LF
Bidens	B. cernua, B. frondosa	SMA
Carex	C. spp.[1]	SM
Carya	C. cordiformis, C. illinoensis	LF, UF
Cephalanthus	C. occidentalis	DMS, SMS
Cornus	C. alternifolia, C. amomum, C. drummondii, C. stolonifera	WMS, SS
Cyperus	C. erythrorhizos, C. esculentus, C. odoratus, C. strigosus	SMA
Decodon	D. verticillatus	DMS, SMS
Echinochloa	E. crusgalli, E. muricata, E. walteri	SMA
Eleocharis	E. obtusa, E. palustris	SMA
Fraxinus	F. nigra, F. pennsylvanica	WS, FF
Juniperus	J. virginiana	CN
Leersia	L. lenticularis, L. oryzoides, L. virginica	WM
Lythrum	L. alatum, L. salicaria	SMP, WM
Nelumbo	N. lutea	RFA
Nuphar	N. lutea, N. variegata	RFA
Nymphaea	N. odorata, N. tuberosa	RFA
Nyssa	N. aquatica, N. sylvatica	WS
Phalaris	P. arundinacea	WM
Phragmites	P. australis	DMP, SMP, WM
Pinus	P. resinosa, P. strobus	CN, PN
Platanus	P. occidentalis	LF
Polygonum	P. spp.	SMA, DMP, SMP, WM
Pontederia	P. cordata	DMP
Populus	P. deltoides	PC, FF
Quercus	Q. spp.	FF, LF, UF
Sagittaria	S. latifolia, S. rigida	DMP, SMP
Salix	S. exigua, S. nigra	SC, SMS, WMS, FF
Sambucus	S. canadensis	WMS
Scirpus	S. spp.	SMP
Solidago	S. spp.	WM
Sparganium	S. eurycarpum	DMP, SMP
Spartina	S. pectinata	WM
Taxodium	T. distichum	WS
Typha	T. angustifolia, T. latifolia	DMP, SMP
Ulmus	U. americana, U. rubra	FF, LF, UF
Zizania	Z. aquatica	DMA

[1] spp. is used when more than four predominant species are present.

www.ingramcontent.com/pod-product-compliance
Lightning Source LLC
Chambersburg PA
CBHW081618170526
45166CB00009B/3021